Understanding the Elements of the Periodic Table™

SILICON

Michael A. Sommers

rosen publishing's
rosen
central®

New York

Pra lui

Published in 2008 by The Rosen Publishing Group, Inc.
29 East 21st Street, New York, NY 10010

First Edition

Library of Congress Cataloging-in-Publication Data

Sommers, Michael A., 1966–
Silicon/Michael A. Sommers.—1st ed.
p. cm.—(Understanding the elements of the periodic table)
Includes bibliographical references and index.
ISBN-13: 978-1-4042-1959-5
ISBN-10: 1-4042-1959-5
1. Silicon. 2. Chemical elements. 3. Periodic law. I. Title.
QD181.S6S59 2008
546'.683—dc22

2007001587

Manufactured in China

On the cover: Silicon's square on the periodic table of elements. Inset: Model of silicon's subatomic structure.

Contents

Introduction

Chemistry is the study of matter and the changes that matter undergoes. (Matter is the name that scientists have given to anything that you can touch, see, feel, or smell—from the air we breathe to the food we eat to the clothes we wear.) Everything on Earth, in our solar system, and in our entire universe that has mass and occupies space is made of matter, including you, your parents, your friends, and all other living things.

The basic building blocks of matter are called elements. Elements are all around us. They are components of all solids, liquids, and gases. Some matter is made up of pure elements (one type of atom), such as the metals iron (Fe), gold (Au), and silver (Ag). You might be out for a walk and come across a rock containing iron, for example. Other matter is made of a combination, or mixture, of elements, known as compounds. For instance, the salt with which you season your food is made up of two elements: sodium (Na) and chlorine (Cl).

When they come into contact with each other, elements may react in different ways, including making completely new types of compounds. Hydrogen (H) and oxygen (O), for example, are two gases. However, when certain amounts of each of these elements come together—creating what is known as a chemical reaction—they form the essential life-giving liquid that we use to drink, bathe, and swim in: water (H_2O).

To date, scientists have identified over 100 different elements. One of the most important and common of these elements is silicon (Si). When most people see or hear the word "silicon," they often think of Silicon Valley, where many high-tech companies have their headquarters. (Silicon Valley got its nickname because silicon is an essential element in computers and other electronic equipment.) They also think of silicone—a class of related compounds containing silicon, oxygen, carbon, and hydrogen. Silicones are used for many purposes, as well as for body implants, such as artificial arms, legs, or breasts. However, silicon is much more than just valleys and implants. It is the second most common element in Earth's crust (after oxygen), making up nearly 28 percent of its mass. Silicon is the seventh most abundant element in the entire universe. There is silicon on the moon and in the sun. Meanwhile, in our day-to-day lives, it is an extremely useful element. Indeed, silicon is present in everything from clay, bricks, glass, and some plastics to toothpaste, hair conditioners, and contact lenses. If it weren't for silicon, you wouldn't have white teeth or soft, shiny hair, let alone your computer or the windows in your bedroom.

Chapter One
The History of Silicon

Silicon has long been an important material. Like gold, silver, iron, and copper (Cu), it has been known and used for thousands of years. However, unlike these elements, which were used in their pure form, the silicon used by our ancestors was found in sand and rocks. Mixed with other elements (oxygen in particular), it was known as silica.

Silica

During the Stone Age (a period ranging from 10,000 to 6,000 years ago, in which humans hadn't yet discovered how to work with metals and therefore made all tools from stone), silicon-based rocks, known as silicates, were made into spears and knives used for hunting and warfare. In North America, one can still find arrowheads made of a common silicate called flint (also known as flintstone) that were used by native peoples 8,000 years ago. A hard blue, gray, or black shiny stone, flint was widely used by prehistoric humans because it splits easily into thin, sharp blades when struck with another hard object. After the Stone Age, it was also discovered that flint produced sparks when struck against steel. When brought close to twigs or branches, flint could be used to start a fire.

These Native American arrowheads are made out of flint. Collecting arrowheads—there are over 200 different types of arrow tips—is a popular hobby in North America.

In fact, the Latin name for silicon—*silex* or *silicis*—means "flint." The ancient Romans, who spoke Latin, discovered new uses for silica. Egyptians had already found that they could crush quartz pebbles and sand (both of which contain silica) into a fine clear powder. Heated at high temperatures and then cooled, this substance became glass. The Romans took glass technology further, blowing the melted silica-based liquid into shapes such as bottles, vases, and containers. From the Egyptians, they also learned how to mix in other elements to create different colors. Adding small amounts of cobalt (Co) ions, for example, created deep blue glass, while the addition of pure gold salts made glass that was ruby red.

Discovery of Silicon

Pure silicon—the element—wasn't discovered until many centuries later. It is believed that the first person to suspect that silicon was an element was the "father of modern chemistry," a French nobleman and scientist named Antoine-Laurent de Lavoisier (1743–1794).

Lavoisier not only named the elements hydrogen and oxygen but also created a list of thirty-three elements. It was Lavoisier who refined the definition of an element as any substance that could not be broken down into a simpler unit. For example, he discovered that water wasn't an element since it could be separated into two other substances: hydrogen and oxygen. (Hydrogen and oxygen are elements because they cannot be broken down any further into atoms of different elements.) In order to determine whether a substance was an element, Lavoisier and other scientists carried out many tests. These included heating and cooling, filtering, and reacting substances with other substances such as acids. Substances that could not be broken down or transformed by any of these means were considered to be elements. This was the case with silicon.

Lavoisier first suspected that silicon was an element when he discovered that silica was a mixture of oxygen and another substance.

An eighteenth-century portrait of Antoine-Laurent de Lavoisier. Lavoisier contributed not only to the history of chemistry but also to the histories of biology, finance, and economics.

"The Father of Modern Chemistry"

Antoine-Laurent de Lavoisier was quite a character. Born in 1743 to a wealthy Parisian family, as a youth he was passionate about science. At the age of twenty-five, he was elected to be a member of the distinguished French Academy of Sciences.

In 1775, he began to work at the Royal Gunpowder Association, where his chemical experiments led to improvements in the production of gunpowder. Indeed, some of Lavoisier's most important experiments examined the nature of burning elements. He demonstrated that burning is a process that involves the reaction of a substance with oxygen. He also wrote what is considered to be the first modern chemistry textbook, *Elementary Treatise of Chemistry* (1789).

Unfortunately, when the French Revolution broke out in 1789, French noblemen were not popular. As a result of his social position, in 1794, Lavoisier was branded a traitor by the new revolutionary government and sentenced to beheading. At the time, he was only fifty-one.

It wasn't until 1823 that Swedish scientist Jöns Jacob Berzelius (1779–1848) isolated pure silicon by reacting a compound containing silicon and filtering out the pure element from other substances formed in the reaction. When purified, he found that silicon was a bluish gray-colored crystal with a metallic shine.

The Periodic Table

Lavoisier was one of the first scientists to create a list of known elements. However, the man who is best known for organizing them into what

In the early nineteenth century, Jöns Jacob Berzelius produced the first accurate table of atomic masses for twenty-eight different elements.

became known as the periodic table of elements was a Russian chemistry professor named Dmitry Mendeleyev (1834–1907). In 1869, Mendeleyev designed a chart that arranged the elements according to their atomic weight, from the lightest to the heaviest. Mendeleyev intended the table to serve as a guide to help his students identify elements and understand their behavior. The periodic table today has evolved significantly from the ones put forth by Mendeleyev and others. However, it is still used extensively by most chemistry students and scientists throughout the world. And of course, over the past 150 years, new information about elements has emerged and many new elements have been discovered.

Reading the Periodic Table

The periodic table is a very useful tool. A one-page chart, it sums up a great deal of information about the elements in a compact manner. The large one-, two-, and three-letter symbols represent the name of each element. Most of them are abbreviations of the element's English name—such as O for oxygen and H for hydrogen. Luckily, silicon is one of the easier elements to identify: its symbol is Si. Others, however, are trickier because their abbreviations come from Latin. Au, for example, stands for *aurum*, the Latin word for "gold." Iron's symbol, Fe, stands for the Latin

The symbol for the element silicon is Si. The number on the left is its atomic number: 14. The number on the right refers to its atomic weight, or mass: 28.

word *ferrum*. Still other elements are named after the place they were discovered or the person who discovered them. Ytterbium (Yb), for example, is named after the town of Ytterby, Sweden. And einsteinium (Es) is named after the famous scientist Albert Einstein.

Reading the periodic table is fairly straightforward. Each element has its own box, which shows the element's symbol as well as its atomic weight and number. The atomic number identifies the element and determines its location on the table. Thus, the elements are listed in increasing atomic number. They start at the top left of the table, beginning with H (hydrogen), which is 1. Families, or groups, of elements (elements that have very similar properties) are listed in vertical columns. Groups use one of two numbering systems. Some periodic tables use a newer system that numbers groups from one to eighteen, moving from left to right. Other tables use an older system, which gives each group a Roman numeral followed by a letter. Silicon is in group 14 or IVA.

Elements in a group usually share some common characteristics, or properties. They often react with other elements or compounds in similar ways. Each group has its own name. Elements in the second column, for example, are known as the alkaline earth metals, while elements in the seventeenth column are halogens. The last—or eighteenth—column consists of elements known as the noble gases. They are unique in that they are the least reactive elements.

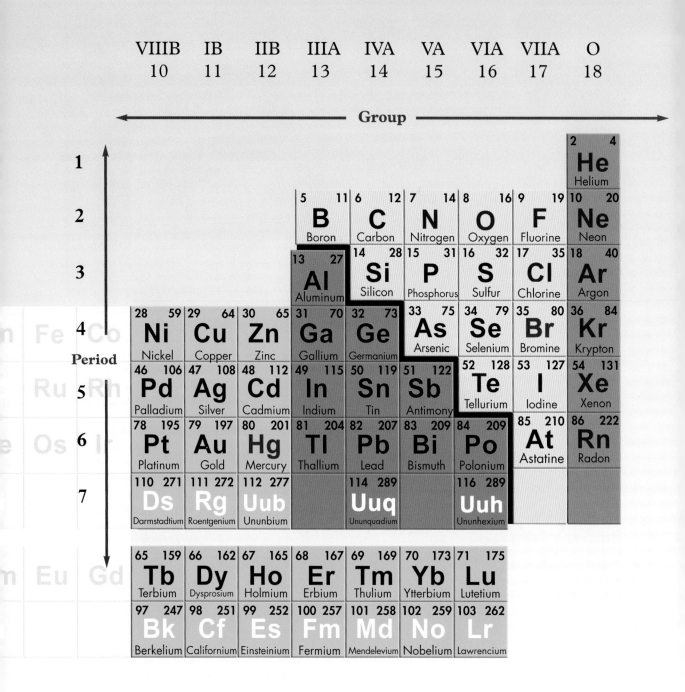

	VIIIB	IB	IIB	IIIA	IVA	VA	VIA	VIIA	O
	10	11	12	13	14	15	16	17	18

Groups 10 (VIIIB) through 18 (O) of the periodic table are displayed above. Visible in bold black is the zigzagging "staircase" that separates metallic elements from semimetals and nonmetals.

Metalloids

As you can tell, a lot of information can be learned just by an element's location on the periodic table. The majority of elements are either metals or nonmetals. Silicon, however, belongs to a group of elements called metalloids, or semimetals. (Silicon is probably the best known of all the metalloids. Arsenic (As), another familiar metalloid, is used in many pesticides and insecticides. In ancient times, however, because it was difficult to detect, arsenic was often used to poison one's enemies.)

Metalloids are elements that are in between metals and nonmetals. They have some characteristics of metals, such as being shiny and hard. However, they also have properties of nonmetals. For instance, unlike gold and silver, metalloids are not malleable, or changeable. This means that if you hammer or try to bend or stretch them, they will usually break.

In the periodic table, metalloids occur along a diagonal line from boron (B) at the top-left to astatine (At) at the bottom-right. Elements to the right of this line are nonmetals; all other elements to the left of the line are metals. Because this line, which separates the metals from the nonmetals, zigzags like a flight of stairs, it is commonly known as the "staircase."

The only other metalloid in the fourteenth column besides silicon is germanium (Ge). Other metalloids are located in the thirteenth, fifteenth, and sixteenth columns. The fact that the metalloids are spread out among various columns means that silicon shares some traits—but not all—with other metalloids. For instance, only some metalloids, most notably silicon, are semiconductors. A semiconductor is a solid through which electricity can flow (although not as easily as with metals and only under certain conditions). Semiconductors are very important for modern technology. They are essential in electrical devices and are used in everything from computers and CD players to cell phones and digital cameras.

Chapter Two
The Atom Explained

In the last chapter, you learned that elements are the building blocks of matter. But what are elements themselves made of? In the eighteenth century, scientists began to suspect that elements were made of tiny particles they called atoms. The word "atom" comes from the Greek term *atamos*, meaning "uncuttable." It refers to the fact that atoms are the basic units of elements and can't be broken down into smaller components that still retain the fundamental properties of that element.

The Atom and Subatomic Particles

In 1803, an English chemist named John Dalton (1766–1844) used the concept of atoms to explain why elements always reacted the way they did. He suggested that each element consists of atoms of a single, unique type and that these atoms could combine to form chemical compounds. Depending on the number of atoms that combined together, different compounds could be created.

Once scientists knew that elements were made up of atoms, they began to dig further to learn what atoms themselves were made of. By the early twentieth century, they had discovered an atom consists of three basic particles that determine its chemical properties: protons, neutrons, and electrons. What makes the elements in the periodic table different from

This is a diagram of a silicon atom. Positively charged protons cluster in the nucleus, along with neutral neutrons. Tiny negatively charged electrons spin around the nucleus in shells.

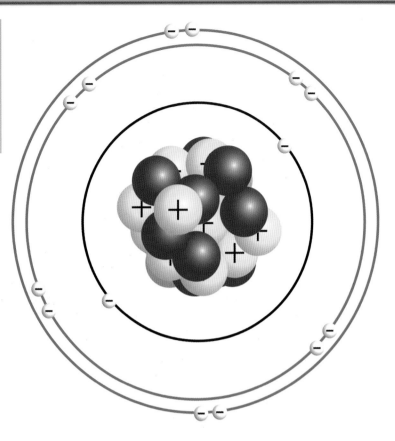

one another is the number of protons each possesses.

Protons

Positively charged protons (represented by the "+" symbol), along with neutrons, make up the center of the atom. Chemists refer to the atom's central core as its nucleus. The number of protons in an atom (the atomic number) is what determines the identity of an individual element—and accounts for its location on the periodic table. For example, if you look at the periodic table, you will see that an atom of silicon, the fourteenth element, always has fourteen protons. If you could take away one proton, you wouldn't have silicon anymore, but instead would have an atom of aluminum (Al), which has thirteen protons. The element with one proton more than silicon is phosphorus (P), which has fifteen protons.

Neutrons

Neutrons are neutral. This means that they have neither a positive nor a negative charge, but a charge of zero (represented by "0"). Neutrons function as a binding force in the nucleus. That is, without neutrons, protons would not stay together because particles with the same charge—whether positive or negative—push each other away. This pushing away is called

Electric Charges

At some time in your life, you have probably received a shock from an electric socket or static electricity. Such shocks are caused by electric charges. An electric charge is produced by the flow of electricity (energy) through particles, namely the tiny particles that make up atoms. Charges can be either positive or negative. Protons have a positive charge (shown by the symbol "+") while electrons have a negative charge (represented by the "–" sign).

repulsion (the opposite of attraction). If you bring the same poles of two magnets together, you will see how repulsion works: the magnets won't stick together. Neutrons are important because they help to cancel out, or "neutralize" protons' repulsive forces, allowing the particles in the nucleus to stay tightly packed together.

Electrons

Electrons have a negative electric charge. As such, their opposite charge to the nucleus helps to hold the electrons in the atom. Together, all of the electrons in a neutral atom create a negative charge that balances the positive charge of its protons. Electrons are much, much smaller and lighter than protons and neutrons, and they contribute little to an atom's weight.

Electrons are located in overlapping regions known as shells. Atoms can have from one to seven shells. The number of electrons in a shell depends on the particular shell. The first shell of an atom, for example, can hold only two electrons, the second shell can hold up to eight, and the third shell can hold up to eighteen. Each shell is a little farther away from the nucleus. Electrons located in inner shells are drawn tightly to the nucleus by the protons' attractive forces. By contrast, the attraction that

Silicon $^{14}_{28}$ Si Snapshot

Chemical Symbol:	Si
Classification:	Metalloid
Properties:	Gray, shiny, odorless, non-malleable, semiconductor
Discovered by:	Jöns Jacob Berzelius
Atomic Number:	14
Average Atomic Weight:	28.0855 atomic mass units (amu)
Protons:	14
Electrons:	14
Neutrons:	14
Density at 68°F (20°C):	2.329 g/cm^3
Melting Point:	2,577°F; 1,414°C; 1,687 K
Boiling Point:	5,909°F; 3,265°C; 3,538 K
Commonly Found:	In compounds with oxygen such as in clay, granite, quartz, sand

binds electrons to the nucleus in outer shells is much weaker. The higher an element's atomic number, the more shells and electrons it will have.

An atom of elemental silicon is neutral and therefore has fourteen protons and fourteen electrons. The electrons are in three shells: the first shell has two electrons, the second has eight, and the third has four.

Bonding

The outermost electrons in an atom are known as valence electrons. These electrons play a special role because they are used to form bonds between atoms to create compounds. Water (H_2O), for example, is a compound in which two hydrogen (H_2) atoms each share electrons with an oxygen (O) atom. In nature, silicon is always found in compounds with other elements. Most commonly, it combines with oxygen to form silica and silicates that make up rocks such as flint and quartz.

Silicon has four valence electrons. This means that silicon would prefer to pick up four electrons to become stable (to have eight electrons). It can get the four missing electrons by sharing electrons from other atoms, which is an example of bonding. Silicon most frequently shares electrons to form four bonds with four different atoms at the same time. This means that silicon can form many different kinds of compounds.

Atomic Weight and Number

If you look at each element on the periodic table, you will notice two numbers. The smaller number is the atomic number. This is the number of protons in an atom of that element. As you saw before, the atomic number of silicon is 14. This is because one atom of silicon has 14 protons. The second, larger number on the table represents the average weight of the element. This weight, or mass, is measured in atomic mass units (amu). The weight is average as opposed to exact because all of an element's

This diagram shows an example of bonding. Here, two oxygen atoms bond with one atom of silicon to form the compound silicon dioxide (SiO_2). Oxygen is the element with which silicon most frequently bonds.

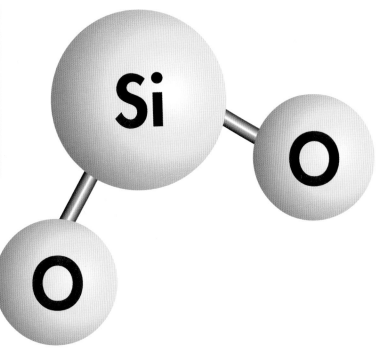

atoms are not always perfectly identical. Often, they naturally occur as isotopes. Isotopes are atoms with the same number of protons, but different numbers of neutrons. Silicon, for instance, has three isotopes. The majority of silicon atoms (92 percent) have fourteen neutrons, and an atomic mass of near 28 amu. However, other silicon atoms have fifteen neutrons, and there are even a few with sixteen neutrons. It is because of these isotopes that silicon's atomic weight is 28.0855 amu.

Chapter Three
Properties of Silicon

I t is important to be able to properly identify an element. For instance, what if gold miners didn't know if the glittery substance they happened upon was a chunk of real gold, worth thousands of dollars, or merely a nearly worthless piece of fool's gold? Similarly, how can you tell if a chunk of rock or grain of sand contains silicon?

Scientists identify an element by the characteristic properties that distinguish it from all other elements. Some properties can be measured using scientific tools and equipment. Others can be observed by looking at the element or by carrying out experiments and watching the changes and reactions that occur. By measuring and observing the appropriate properties of any sample of matter—whether a solid, liquid, or gas—you can tell what kinds of element it contains.

Observing Properties

At room temperature, pure silicon is a solid that exists in two forms. One form is made up of shiny black or blue-gray crystals. The other is a dull brown powder. Forms of silicon look different depending on the way their atoms are packed together. In the case of the shiny crystals, silicon atoms join together in an identical, regular pattern. In the dull brown powder, the tiny crystals are so small that they appear powderlike.

The shiny black lumps of silicon *(top)* are samples of the element in its crystal form. Silicon also occurs in powder form *(bottom)*. The powder is actually made up of very tiny clusters of crystals.

Measuring Properties

If you were to run tests on silicon samples, you would discover some more interesting properties about the element. For instance, you would have to turn up the heat extremely high in order to get silicon to melt from a solid to a liquid. This is because in both crystal and powder forms, silicon melts at a temperature of 2,577 degrees Fahrenheit (1,414 degrees Celsius). Its boiling point—the temperature at which liquid silicon becomes a gas—is even higher: 5,252°F (2,900°C).

At room temperature, pure silicon has a density of 2.329 grams per cubic centimeter. This means that a sample of silicon weighs 2.329 times more than an identical volume of water. As a metalloid, silicon is hard—but not too hard. On the Mohs' scale (a scale that ranges from 1 to 10, ranking minerals' hardness or softness based on how difficult they are to scratch), silicon rates a 6.5. In comparison, a diamond, the hardest naturally occurring material, is a 10.

A Semiconductor

One of silicon's claims to fame is its role as a semiconductor. It is this property that makes it so valuable for use in computer and electronic devices. Metals are naturally good at conducting electricity. This is because some of the electrons in metals are not tightly tied to their individual atoms. Instead, they can roam fairly freely throughout the metallic element, allowing the transfer of electricity. For this reason, metals such as copper or aluminum are used in electronic wiring and devices such as televisions and sound systems.

In contrast, electrons in nonmetals are bound much more tightly to their individual atom. Nonmetals are typically found in compounds with filled valence shells. Therefore, electricity cannot travel through these substances under normal conditions. Thus, nonmetals are almost always

This close-up photograph shows a silicon chip inside a computer. Silicon chips are an essential part of computer hardware and for many other electronic devices commonly used in the twenty-first century.

insulators, not allowing the flow of electricity. Insulating compounds made from nonmetals are often used in home appliances to prevent us from getting electrical shocks.

Because silicon is a metalloid, it has unique properties. At low temperatures, silicon functions like a typical nonmetallic insulator. The four electrons in its outer shell are bound tightly to the atom. However, if silicon's valence electrons are given extra energy—in the form of heat, for example—they can break free from the bonds and orbit around, carrying electricity. This makes silicon a semiconductor and means that the electrical flow through silicon can be controlled as needed. Silicon can carry electricity, but its insulating properties, especially when mixed with oxygen as silicon dioxide (SiO_2), prevent the silicon from getting too hot or the electrical current from growing too strong.

How Silicon Reacts

Pure elemental silicon does not usually occur naturally on Earth. The only exceptions are rare crystals that are sometimes found along with gold or in eruptions of volcanic material. Silicon also hardly ever reacts with other elements. No strong acids have an effect on silicon. The only elements with which silicon reacts are halogens such as fluorine (F) and chlorine, and some alkalis, a group of chemicals that includes sodium hydroxide (NaOH). The most important element with which silicon reacts is oxygen. When silicon combines with oxygen, the result is silicon dioxide (SiO_2), also known as silica.

Silicon is present in large quantities on Earth, making up nearly 28 percent of Earth's crust. Silicon is found combined with oxygen as silica, and as thousands of different types of silicates. Sand, clay, granite, agate, quartz, rock crystal, flint, jasper, and opal are some of the mineral forms in which silicon appears mixed with oxygen and other elements.

Silicon is present in a variety of precious and semiprecious stones, such as this beautiful shiny chunk of agate *(above)* and this glittering opal *(right)*.

Silicon is present in certain types of meteorites (rocks that fall to Earth from outer space) and moon rocks. It is also a major ingredient in tektite, a natural form of glass. Meanwhile, olivine is a very simple silicate that combines silicon, oxygen, and magnesium (Mg) or iron ions. Olivine is a hard, glasslike, silicon-containing mineral that makes up most of Earth's mantle, the region between the planet's core and its surface, or crust.

You and Silicon

Silicon is an essential part of many living creatures, probably including human beings. It is thought to be an important ingredient in our bones, arteries, tendons, and skin (making it elastic). Scientists believe that silicon may help to strengthen these tissues.

Life Source

Silicon is directly below carbon (C) on the periodic table. They are in the same group of elements so they share many similarities. As you might already know, all life on Earth depends on carbon. Carbon is so important because it is capable of forming millions of different compounds, including the cells that give life to plants, animals, and humans.

Like silicon, carbon is able to bond with so many other elements because of the four valence electrons it has in its outer shell. The many similarities between carbon and silicon have made some scientists wonder whether life based on silicon atoms is possible. There are also some significant differences between silicon and carbon, however, and to date, there are no known forms of life that rely entirely on silicon. Still, the notion of living creatures—including humans—that evolved out of silicates such as clay is popular with science fiction writers.

Chapter Four
Silicon Compounds

Silicon in nature is essentially always found in the form of a compound. The most common of these compounds is silicon dioxide (SiO_2), or silica, a compound of silicon with oxygen. If you took a walk out in nature, no matter where in the world you happened to be, almost all of the rocks you'd encounter would contain silicon and oxygen, and frequently also ions of calcium (Ca), magnesium, or chlorine.

Quartz

A very common form of silica is quartz. Pure quartz consists of clear crystals made up of regular patterns of silicon and oxygen. It is the purest form of silica. When quartz contains some other elements, it can take on beautiful colors. Quartz with traces of iron ions becomes a purplish-lilac stone called amethyst, a precious gemstone. Another beautiful and greatly valued compound related to quartz also has ions of chromium (Cr), beryllium (Be), and aluminum: it is an emerald!

Have you ever wondered why some wristwatches say "quartz" on them? They actually contain tiny quartz crystals. The watch contains a battery to produce an electrical current. When the current passes through the tiny crystal, it causes the crystal to vibrate. It is these small but steady vibrations

Purple amethyst is a type of quartz containing silicon. Amethyst has long been a popular gemstone. It was even used in ancient Greek and Roman times.

that make clocks and watches tick with regularity and accuracy, thus allowing us to tell time.

Sand

When quartz and other silica-based rocks are ground down, they become a powdery substance we call sand. Actually, most sand is tiny grains of silica that have been ground down over time by severe weather and by the action of sea waves crashing down upon the shore.

Aside from being nice to lie on or make castles with, sand is incredibly useful. It is the most important ingredient in making glass. In fact, glass is essentially liquid sand, which after being melted, is solidified into new

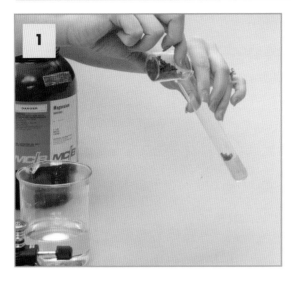

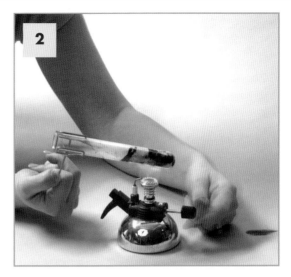

This experiment shows that pure silicon can be separated from sand. *(1)* Sand (which contains silica) and magnesium are mixed together in a test tube. *(2)* The mixture is heated until it glows and the mixture reacts. *(3)* The test-tube contents are poured into a beaker containing hydrochloric acid, which *(4)* dissolves the by-products produced in the reaction. *(5)* After the acid is poured off, gray silicon powder is left at the bottom of the beaker, along with some unreacted sand.

shapes. Silicon is the most important element in all kinds of glassware, from drinking cups, soft-drink bottles, and windows to heat-resistant baking dishes and fine crystal chandeliers.

Clay

Clay is also usually made up of silicon- and oxygen-containing compounds that exist as many tiny flat particles that stick together when water is added to them. Clay can be molded into an endless variety of shapes. When it is heated and the water within it evaporates, it dries and hardens. Silicon is therefore an essential ingredient in all types of ceramics and pottery, as well as in the clay bricks that are used to construct homes and buildings. Moreover, the cement used to hold bricks together also contains clay.

Silicon Carbide

Silicon carbide (SiC) is a compound made by heating pure sand, containing silica, and carbon in a furnace at a very high temperature. When cooled to a solid, the resulting compound is one of the hardest artificial substances in existence. Silicon carbide is used in sandpapers and as a coating for saws. It makes blades so hard that they can cut through most rocks and gemstones.

Silicones

One of the most useful classes of silicon compounds is referred to as silicone. Silicones are compounds that vary in consistency from soft, gel-like liquids and rubbery solids to hard, plastic-like solids. Silicones are made from silicon, oxygen, carbon, and hydrogen atoms. In fact, one of the first silicones, invented in the early twentieth century, was used as a replacement for the natural rubber covers that insulated electrical cables.

Silicones come in a variety of forms and are used to make a wide range of products, including artificial limbs, sunscreen, contact lenses, and waterproof rubber boots.

U.S. astronaut Neil Armstrong—and his silicon soles—made history when they left the first human footprint on the surface of the moon.

Silicones are now used in an enormous variety of products, including space suits, waterproof clothing, paints, suntan lotions, kitchen spatulas, contact lenses, and the rubbery gels used to seal leaky doors and windows and repair punctures in bicycle tires.

Silicones are unaffected by high temperatures and are extremely nonreactive (they don't bond easily with other substances). This makes them ideal for products that need to be sterilized to kill germs and that won't be affected when they come into contact with other chemicals. For these reasons, silicone is an ideal material with which to make artificial limbs (particularly fingers and toes) and breast implants.

There are many other useful products that contain silicone. Since silicone is heat resistant, it is efficient for fighting fires. Silicone foams are often used to seal fire doors and to put out flames. Silicones are also used as waterproof seals in everything from windbreakers and rain boots to cement and gels for fixing leaks. Silicone even made history. In 1969, when American astronaut Neil Armstrong was the first human to set foot on the moon, the footstep that he left was made with a boot sole that was made of lightweight silicone.

Chapter Five
Silicon in Our Lives

I t is not an exaggeration to say that we are living in a silicon age. Silicon is an essential element in our lives. In particular, since it is a critical element in computers and other electronic equipment, silicon is responsible for our experiencing a "technological revolution." In fact, for this reason, the region of California (near San Francisco) where many high-tech companies have their headquarters is referred to as Silicon Valley. You have probably heard of companies such as Yahoo!, Google, eBay, Hewlett-Packard, and Apple. All of these highly profitable groups depend—directly or indirectly—on silicon. Without silicon, the countless electronic innovations that we take for granted wouldn't exist as we know them.

Purifying Silicon

Since pure silicon rarely occurs in nature, the only way to get pure elemental silicon is by separating it from its compounds. Today, the majority of pure silicon comes from heating a type of quartz sand with carbon in a very hot furnace at a temperature over 5,400°F (3,000°C). At this temperature, the carbon reacts with and removes the oxygen from the silica as carbon monoxide (CO) gas. This leaves liquid silicon that is between 98 and 99 percent pure.

The city of San Jose, California, lies in the center of the region known as Silicon Valley, which is famous for its large number of high-tech companies.

Most of the remaining 1 to 2 percent of impurities can be removed through a process called distillation. The silicon liquid is heated to its boiling point. While the silicon becomes a gas, the leftover impurities (which boil at a higher temperature) remain behind. Once the silicon gas cools to a solid, you have elemental silicon that is quite pure. Further purification can be done by heating one end of a rod of solid silicon with a powerful electric heater that slowly moves along the length of the rod. The impurities therefore collect at one end of the rod, which is then cut off and thrown away. What remains can be up to 99.999999 percent pure. This purifying method, known as zone refining, is done on a large scale by industries all over the world that use pure silicon to make computers and other electronic devices.

Silicon Chips

Today, it is hard to believe that the earliest computers took up the space of entire large rooms. How did computers come to be the size of a laptop? It was due to a miraculous invention called the silicon chip. Silicon chips are tiny devices that look like thin wafers. They are actually tiny fragments cut from pure silicon crystals. Pure silicon is melted into liquid form, which is then allowed to harden into a silicon crystal.

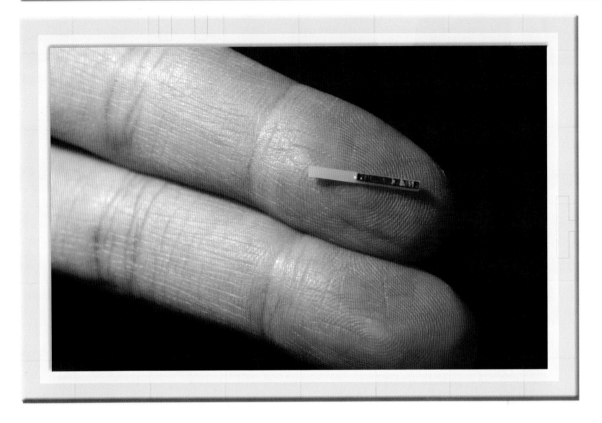

This tiny strip is a silicon laser chip that is an important element in the production of high-speed, low-cost computers.

Because silicon is so inexpensive and abundant—and is easily made into a semiconductor—it is ideal for making small and powerful devices. The size of all those devices is a big difference from the room-sized computers of sixty years ago.

Silicosis

Silicon and silicon compounds have given us enormous benefits, but there is a potential drawback as well. Silica is the cause of a disease called silicosis, which historically has killed hundreds of people in North America every year. Miners, stone cutters, glass manufacturers, pottery makers,

Quite a Mouthful

The full name for silicosis is pneumonoultramicroscopicsili-covolcanoconiosis! A mouthful to pronounce, at forty-five letters, it is the longest word in the English language!

and other people who regularly work with silica and breathe the dust into their lungs are those who usually get the disease.

It can take years for silicosis to develop. It then can be difficult to treat. It usually attacks the lungs, causing swelling and scarring. Symptoms usually include shortness of breath, fevers from secondary infections, and chronic cough. Fortunately, it is relatively easy to prevent silicosis in the first place. Wearing a mask and working in a place with good ventilation provides protection from breathing in this harmful dust.

Living in the Silicon Age

It is impossible to imagine a day in your life without silicon. Consider this more or less normal day: You wake up to the ringing of your alarm clock, whose hard casing contains silicone and whose hands' regular ticking is due to a tiny particle of quartz, or silica. Realizing you're late for school, you rush into the bathroom and take a shower. The shampoo and conditioner contain silicon compounds that will help to leave your hair soft and shiny.

Back in your bedroom, you get dressed (noticing how soft your jeans are due to the silicone compound in the fabric softener). Then you go downstairs to the kitchen where your dad is serving muffins that were baked on parchment paper, which contains a silicon compound to prevent sticking. While you're eating, you gaze out the window (recalling that glass contains silicon dioxide) and notice that your mother's rock garden

DIRECTIONS: After shampooing, massage conditioner through hair and rinse thoroughly after 1 to 3 minutes. May be used every time you wash your hair.

INGREDIENTS: Water (Aqua), Cetearyl Alcohol, Glycerin, Cetrimonium Chloride, Fragrance (Parfum), Dimethiconol, Quaternium-18, Potassium Chloride, Disodium EDTA, Propylene Glycol, DMDM Hydantoin, TEA-Dodecylbenzenesulfonate, Lysine HCl, Silk Amino Acids [Alanine, Glycine, Serine, Arginine, Isoleucine, Cystine, Histidine, Glutamic Acid], Borago Officinalis Seed Oil (Borago Officinalis)[Palmitic Acid, Stearic Acid, Linoleic Acid, Oleic Acid, Eicosenoic Acid], Methylchloroisothiazolinone, Methylisothiazolinone.

Silicones in shampoos and conditioners make freshly washed hair appear shiny and feel soft. Dimethiconol *(listed on a conditioner bottle, above)* is a compound that contains silicon atoms.

is looking especially pretty as the early morning sun glints off the quartz embedded in the rocks.

The cement sidewalk on which you walk to school contains silica. So do the bricks of all the houses in your neighborhood, as well as those of your school building. You notice that your neighbor's home, which is fueled by solar energy, captures this energy with solar panels containing semiconductor cells made of silicon. Later, back at home, the house smells fresh and clean. Your dad scrubbed the floors using cleaning products containing silicones. He also painted the kitchen bright green using a paint with a protective silica coating. Your mother left you a Post-It note—whose sticky substance contains silicon—reminding you to record her favorite television show: both the TV and DVD player's technologies depend upon silicon.

After dinner, you turn on your computer to e-mail some pictures to your friends, knowing that neither your laptop, your digital camera, nor your favorite computer games would work if it wasn't for the silicon chips installed inside them. Before you turn in for the night, you realize that if it weren't for silicon, the world as you know it wouldn't exist!

The Periodic Table of Elements

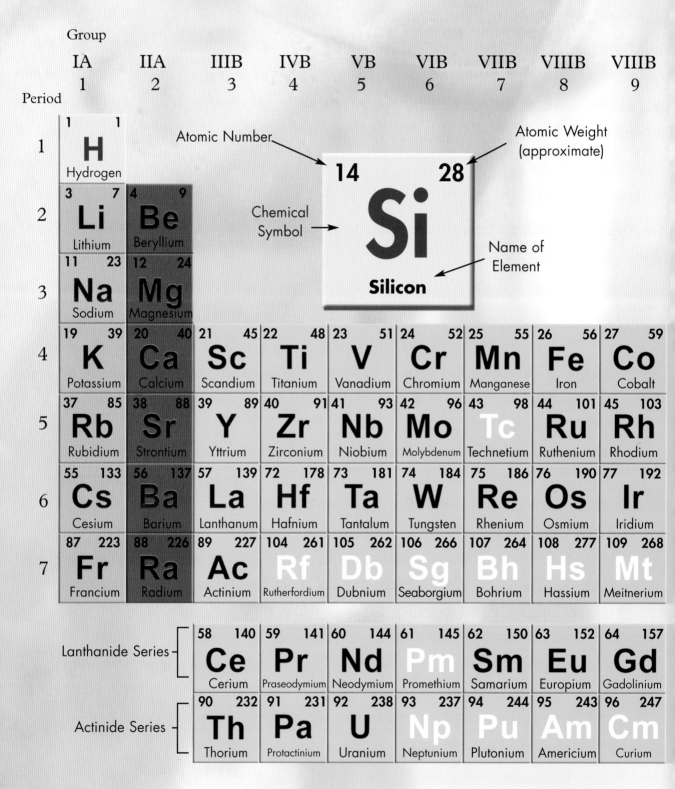

Group

IA	IIA	IIIB	IVB	VB	VIB	VIIB	VIIIB	VIIIB
1	2	3	4	5	6	7	8	9

Period

Atomic Number

Atomic Weight (approximate)

Chemical Symbol

14   28

Si

Name of Element

Silicon

Period								
1	**1 1** H Hydrogen							
2	**3 7** Li Lithium	**4 9** Be Beryllium						
3	**11 23** Na Sodium	**12 24** Mg Magnesium						
4	**19 39** K Potassium	**20 40** Ca Calcium	**21 45** Sc Scandium	**22 48** Ti Titanium	**23 51** V Vanadium	**24 52** Cr Chromium	**25 55** Mn Manganese	**26 56** Fe Iron	**27 59** Co Cobalt
5	**37 85** Rb Rubidium	**38 88** Sr Strontium	**39 89** Y Yttrium	**40 91** Zr Zirconium	**41 93** Nb Niobium	**42 96** Mo Molybdenum	**43 98** Tc Technetium	**44 101** Ru Ruthenium	**45 103** Rh Rhodium
6	**55 133** Cs Cesium	**56 137** Ba Barium	**57 139** La Lanthanum	**72 178** Hf Hafnium	**73 181** Ta Tantalum	**74 184** W Tungsten	**75 186** Re Rhenium	**76 190** Os Osmium	**77 192** Ir Iridium
7	**87 223** Fr Francium	**88 226** Ra Radium	**89 227** Ac Actinium	**104 261** Rf Rutherfordium	**105 262** Db Dubnium	**106 266** Sg Seaborgium	**107 264** Bh Bohrium	**108 277** Hs Hassium	**109 268** Mt Meitnerium

Lanthanide Series

58 140 Ce Cerium	**59 141** Pr Praseodymium	**60 144** Nd Neodymium	**61 145** Pm Promethium	**62 150** Sm Samarium	**63 152** Eu Europium	**64 157** Gd Gadolinium

Actinide Series

90 232 Th Thorium	**91 231** Pa Protactinium	**92 238** U Uranium	**93 237** Np Neptunium	**94 244** Pu Plutonium	**95 243** Am Americium	**96 247** Cm Curium

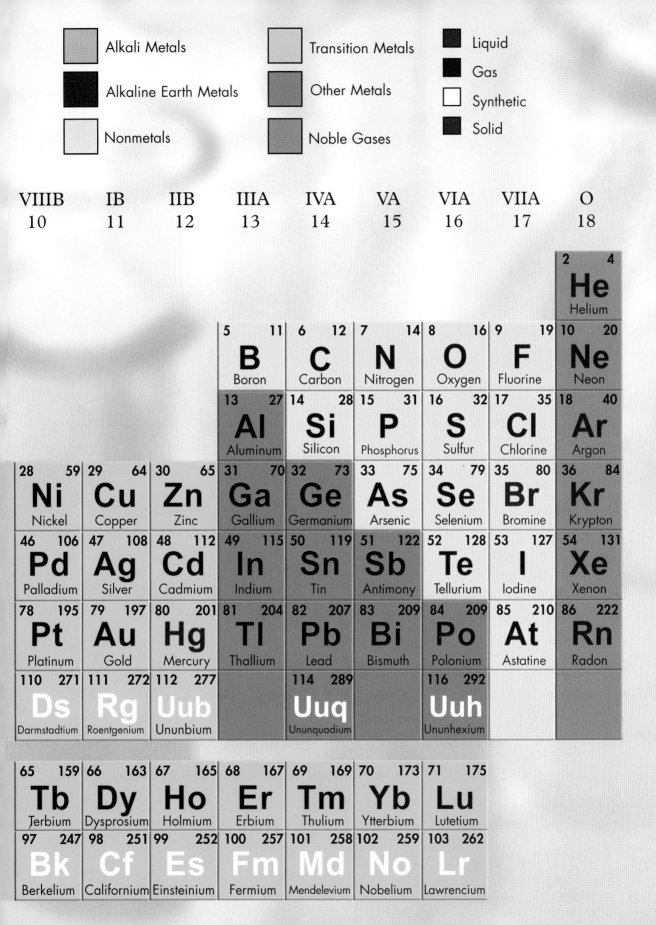

atom The smallest particle of an element that retains the properties of that element.

atomic mass The mass or weight of one atom.

atomic number The number of protons in an atom. This number also identifies the element's place on the periodic table.

bond An attraction between two or more atoms or ions that holds them together.

chemistry The study of the properties of matter and how matter changes.

compound A substance made from two or more atoms.

conductor A substance (usually metallic) that carries electricity (electrons).

crystal A solid in which atoms, ions, or molecules are packed in a regular three-dimensional pattern.

density A substance's mass per unit volume.

electron A negatively charged particle outside the nucleus of an atom.

element A pure substance made from one type of atom.

insulator A substance (usually nonmetallic) that is a poor transmitter of electricity (electrons).

isotope Atom of an element that has the same number of protons and electrons but a different number of neutrons.

matter Any substance that has mass (weight) and occupies space.

Mohs' scale A scale, ranging in value from 1 to 10, that measures the softness or hardness of a mineral.

neutron A tiny particle with no electrical charge found in the nucleus of an atom.

nucleus An atom's center or core, which contains protons and neutrons.

periodic table A chart of all the chemical elements arranged according to their atomic number and grouped by similar properties.

proton A tiny particle with a positive charge located in the nucleus of an atom.

refine A process to isolate and purify to form, for example, a pure element.

semiconductor A substance through which electricity flows more easily under certain conditions than through an insulator, but less easily than through a conductor.

For More Information

American Chemical Society
1155 Sixteenth Street NW
Washington, DC 20036
(800) 227-5558 or (202) 872-4600
Web site: http://www.acs.org

Center for Science & Engineering Education (CSEE)
Lawrence Berkeley National Laboratory
1 Cyclotron Road, MS 7R0222
Berkeley, CA 94720
(510) 486-5511
Web site: http://csee.lbl.gov

Environmental Literary Council
1625 K Street NW, Suite 1020
Washington, DC 20006
(202) 296-0390
Web site: http://www.enviroliteracy.org

International Union of Pure and Applied Chemistry (IUPAC)
IUPAC Secretariat
P.O. Box 13757
Research Triangle Park, NC 27709-3757
(919) 485-8700
Web site: http://www.iupac.org

Mineral Information Institute
505 Violet Street
Golden, CO 80401
(303) 277-9190
Web site: http://www.mii.org

Web Sites

Due to the changing nature of Internet links, Rosen Publishing has developed an online list of Web sites related to the subject of this book. This site is updated regularly. Please use this link to access the list:

http://www.rosenlinks.com/uept/sili

Cooper, Christopher. *Matter* (Eyewitness Books). New York, NY: DK Children's Books, 1999.

Kjelle, Marylou Morano. *Antoine Lavoisier: Father of Chemistry* (Uncharted, Unexplored, and Unexplained). Hockessin, DE: Mitchell Lane Publishers, 2004.

Miller, Ron. *The Elements: What You Really Want to Know.* Minneapolis, MN: Twenty-First Century Books, 2006.

Newmark, Ann. *Chemistry* (Eyewitness Books). New York, NY: DK Children's Books, 2005.

Parker, Steve. *Chemicals & Change* (Science View). New York, NY: Chelsea House Publishers, 2004.

Saunders, Nigel. *Carbon and the Elements of Group 14* (The Periodic Table). Chicago, IL: Heinemann Library, 2003.

Stwertka, Albert. *A Guide to the Elements.* 2nd ed. New York, NY: Oxford University Press, Inc., 2002.

Thomas, Jens. *Silicon* (The Elements). New York, NY: Benchmark Books, 2001.

Yount, Lisa. *Antoine Lavoisier: Founder of Modern Chemistry* (Great Minds of Science). Updated ed. Berkeley Heights, NJ: Enslow Publishers, Inc., 2001.

Brandolini, Anita. *Fizz, Bubble, and Flash! Element Explorations and Atom Adventures for Hands-on Science Fun!* Charlotte, VT: Williamson Publishing, 2003.

Chem4kids. Retrieved December 2006 (http://www.chem4kids.com).

Environmental Literacy Council. "Silicon." Retrieved December 2006 (http://www.enviroliteracy.org/article.php/1014.html).

Frank, David V., John G. Little, and Steve Miller. *Chemical Interactions* (Prentice Hall Science Explorer). Upper Saddle River, NJ: Pearson Prentice Hall, 2005.

Mineral Information Institute. "Silicon or Silica." Retrieved December 2006 (http://www.mii.org/Minerals/photosil.html).

PBS.org. Frontline: Breast Implants on Trial. "Silicone: A Brief History." Retrieved December 2006 (http://www.pbs.org/wgbh/pages/frontline/implants/corp/history.html).

Thomas, Jens. *Silicon.* New York, NY: Benchmark Books, 2002.

About the Author

Michael A. Sommers was born in Texas and raised in Canada. After earning a bachelor's degree in English literature at McGill University, in Montreal, Canada, he went on to complete a master's degree in history and civilizations from the Ecole des Hautes Etudes en Sciences Sociales, in Paris, France. For the last fifteen years, he has worked as a writer and photographer.

Photo Credits

P. 7 © www.istockphoto.com/George Bailey; p. 8 © Mary Evans/Photo Researchers, Inc.; p. 10 © Sheila Terry/Photo Researchers, Inc.; p. 21 (top) © Andrew Lambert Photography/Photo Researchers, Inc.; p. 21 (bottom) © Theodore W. Gray; p. 23 © www.istockphoto.com/Larry Treat; p. 25 (top) © www.istockphoto.com/Hubert Cola; p. 25 (bottom) © www.istockphoto.com/Alice Millikan; p. 28 © www.istockphoto.com/ Emrah Turudu; p. 29 Mark Golebiowski; p. 31 (top, left) © Yoshikazu Tsuno/AFP/Getty Images; p. 31 (top, right) © www.istockphoto.com/ Jan Rihak; p. 31 (bottom, left) © www.istockphoto.com/Judy Allan; p. 31 (bottom, right) © www.istockphoto.com/Ken Hurst; p. 32 © NASA/ Newsmakers/Getty Images; p. 34 © David McNew/Newsmakers/Getty Images; p. 35 © Justin Sullivan/Getty Images; p. 37 (left) Cindy Reiman; p. 37 (right) © www.istockphoto.com/Tammy Bryngelson.

Special thanks to Jenny Ingber, high school chemistry teacher, Region 9 Schools, New York, New York, for her assistance in executing the science experiments illustrated in this book.

Designer: Tahara Anderson; **Photo Researcher:** Amy Feinberg